Supporting Sustainable Rural Communities

Partnership for Sustainable Communities

In collaboration with the U.S. Department of Agriculture

The Partnership for Sustainable Communities

On June 16, 2009, U.S. Department of Transportation (DOT) Secretary Ray LaHood, U.S. Department of Housing and Urban Development (HUD) Secretary Shaun Donovan, and U.S. Environmental Protection Agency (EPA) Administrator Lisa P. Jackson announced the formation of the interagency Partnership for Sustainable Communities. This action marked a fundamental shift in the way the federal government structures its transportation, housing, and environmental policies, programs, and spending, and Americans are already seeing the impacts. The three agencies are working together to support urban, suburban, and rural communities' efforts to attract economic growth, expand housing and transportation choices, protect their air and water, and provide the type of development residents want.

Through the Partnership and guided by six Livability Principles (below), HUD, DOT, and EPA are coordinating investments and aligning policies to support sustainable communities—places that provide homes working families can afford, reliable and economical transportation options, shopping and other daily needs close to where people live, and vibrant and healthy neighborhoods that attract young people and businesses.

The Partnership breaks down the traditional silos of housing, transportation, and environmental policy to consider these issues as they exist in the real world—inextricably connected. Coordinating federal investments yields better results for communities and uses taxpayer money more efficiently by meeting multiple economic, environmental, and community objectives with each dollar spent. For example, investing in the revitalization of a town's Main Street can spur business development, catalyze the renovation of historic structures, save taxpayer dollars by avoiding the need for new streets and water infrastructure, encourage healthy walking and bicycling, and give residents transportation choices that can save them money and reduce air pollution.

Partnership for Sustainable Communities Guiding Livability Principles

Provide more transportation choices. Develop safe, reliable, and economical transportation choices to decrease household transportation costs, reduce our nation's dependence on foreign oil, improve air quality, reduce greenhouse gas emissions, and promote public health.

Promote equitable, affordable housing. Expand location- and energy-efficient housing choices for people of all ages, incomes, races, and ethnicities to increase mobility and lower the combined cost of housing and transportation.

Enhance economic competitiveness. Improve economic competitiveness through reliable and timely access to employment centers, educational opportunities, services and other basic needs by workers, as well as expanded business access to markets.

Support existing communities. Target federal funding toward existing communities—through strategies like transit-oriented, mixed-use development and land recycling—to increase community revitalization and the efficiency of public works investments and safeguard rural landscapes.

Coordinate and leverage federal policies and investment. Align federal policies and funding to remove barriers to collaboration, leverage funding, and increase the accountability and effectiveness of all levels of government to plan for future growth, including making smart energy choices such as locally generated renewable energy.

Value communities and neighborhoods. Enhance the unique characteristics of all communities by investing in healthy, safe, and walkable neighborhoods—rural, urban, or suburban.

Table of Contents

Executive Summary.. 1

Introduction 2

How the Livability Principles Support Rural Communities 5

HUD, DOT, EPA, and USDA Programs At Work in Rural Communities 8

Performance Measures for Success.. 15

Conclusion and Next Steps for the Partnership. 20

Appendix A: Case Studies of Federal Support for Sustainable Rural Communities.

Appendix B: Partnership for Sustainable Communities Charge to the Rural Work Group

Appendix C: Rural Work Group Members

Executive Summary

Rural communities across America are working to strengthen their economies, provide better quality of life to residents, and build on assets such as traditional main streets, agricultural and working lands, and natural resources. The Partnership for Sustainable Communities, in collaboration with the U.S. Department of Agriculture (USDA), established a Rural Work Group to reinforce these initiatives and ensure that the four agencies' spending, policies, and programs support rural communities' efforts to be economically vibrant and environmentally sustainable. This report summarizes the Rural Work Group's findings and creates a framework for the Partnership's future work with rural communities.

The report includes the following sections:

- **How the Livability Principles Support Rural Communities**: The Livability Principles that guide the Partnership provide a useful policy framework for supporting sustainable rural communities, making a critical connection between economic competitiveness, agricultural and natural land preservation, the leveraging of existing infrastructure, and quality of life. This section articulates how the Livability Principles apply in the rural context.

- **HUD, DOT, EPA, and USDA Programs at Work in Rural Communities**: HUD, DOT, EPA, and USDA each make significant investments and implement policies in rural America and are positioned to support sustainable community development in rural communities and regions. This section provides examples of federal programs at work in rural communities.

- **Performance Measures for Success**: Communities of all sizes are using performance measurement to understand the impacts of their programs, policies, and investments. This section identifies a sample set of performance measures tailored to the rural context and organized under four broad goals: promoting rural prosperity, supporting vibrant and thriving rural communities, expanding transportation choices, and providing affordable housing opportunities.

- **Conclusion and Next Steps for the Partnership**: This section outlines a set of next steps the Partnership agencies are considering to support the efforts of rural communities and small towns to invest in a sustainable future.

- **Case Studies of Federal Support for Sustainable Rural Communities**: Across the country, rural communities are strengthening their existing neighborhoods, providing more transportation and housing choices, and promoting economic development that complements their rural character. Appendix A provides some examples of how federal agencies are supporting these efforts.

Introduction

Background

Rural communities across America are working to strengthen their economies, provide better quality of life to residents, and build on assets such as traditional main streets, agricultural and working lands, and natural amenities and resources. The Partnership for Sustainable Communities—made up of the U.S. Department of Housing and Urban Development (HUD), the U.S. Department of Transportation (DOT), and the U.S. Environmental Protection Agency (EPA)—is coordinating with the U.S. Department of Agriculture (USDA) to reinforce these initiatives and ensure that the four agencies' spending, policies, and programs support rural communities' efforts to be economically vibrant and environmentally sustainable.

HUD, DOT, EPA, and USDA already make significant investments and implement policies in rural America through mechanisms such as USDA Rural Development loans and grants, HUD's State and Small Cities Community Development Block Grant and Housing Choice Voucher programs, DOT's rural transit expenditures, and EPA's clean water and drinking water state revolving funds. Additionally, the Partnership supports community and regional planning efforts in rural areas. For example, in 2010, HUD awarded $28 million in Sustainable Communities Regional Planning Grants to regions with populations less than 500,000 and $15 million in Community Challenge Planning Grants to rural places with fewer than 200,000 people. The 13,000-resident City of Glens Falls, New York, for instance, received funding to develop a strategy to provide affordable workforce housing downtown, identify vacant properties for infill development, and amend its zoning ordinance to increase energy efficiency. The Housing Authority of Randolph County, with a population of 28,000, received a grant to develop a county-wide plan that identifies areas for farmland preservation, assesses opportunities for expanding bus service, and increases pedestrian and bike connectivity.

Strengthening federal support for rural communities by coordinating and aligning these programs is a key Partnership goal. In August 2010, the Partnership established a Rural Work Group comprised of staff from HUD, DOT, EPA, and USDA to guide its approach to rural sustainable communities. This report, which summarizes the work group's efforts, explores how the Partnership can contribute to more resilient economies, healthy environments, and quality of life in rural America. It also sets out a framework for the Partnership's future work with rural communities.

The report includes the following sections:

- **How the Livability Principles Support Rural Communities**: This section articulates how the six Livability Principles that guide the Partnership apply in the rural context.
- **HUD, DOT, EPA, and USDA Programs at Work in Rural Communities**: This section describes programs at work in rural communities at each of the four agencies.
- **Performance Measures for Success**: This section identifies performance measures that local, regional, and federal policymakers can use to assess the effectiveness of sustainable communities approaches in small towns and rural places.

- **Conclusion and Next Steps for the Partnership**: This section outlines a set of steps the Partnership agencies can take to support the efforts of rural communities and small towns as they invest in a sustainable future.
- **Case Studies of Federal Support for Sustainable Rural Communities:** Appendix A describes rural communities that have successfully implemented sustainable communities approaches with assistance from federal agencies.

The Rural Context

Rural is difficult to define. A rural community in a relatively high-population state can look dramatically different from a similarly sized rural community in a less populous state. One definition cited by the USDA Economic Research Service describes rural areas as nonmetropolitan counties. By this definition, nearly two-thirds of the nation's 3,142 counties are rural, and rural communities comprise 17 percent of the population (49 million people) and about 80 percent of the country's total land area.[1] However, these statistics, while important, do not describe the interaction between communities and their surrounding landscapes that is so integral to understanding the challenges and opportunities in rural areas.

From a land use and development perspective, rural America includes towns and small cities as well as working lands, or lands that are managed for economic value such as farms, prairies, forests, and rangelands. Historically, rural land was often used for the production and extraction of resources. Towns were developed at transportation hubs—rail stations, river ports, major crossroads—providing the places where agricultural or natural resources could be traded or shipped. Many rural communities were built around main commercial streets and relatively compact, walkable neighborhoods, along with valuable infrastructure that served their civic, cultural, and social needs. The working lands surrounding the towns often provided the reason for their existence, and continue to do so in many places. The rural landscape is more than attractive vistas—it is integral to the social and economic life of the community.[2]

Today, rural communities face an array of challenges. Resource-based economies are vulnerable to the impacts of commodity prices, technological changes, land value dynamics, and other market influences. Some communities whose economies are contracting are experiencing unemployment, poverty, population loss, the aging of their workforces, and increasing demands for social services with fewer dollars to pay for them. In some rural areas, these are not new trends, but generations-old issues. Additionally, residents of remote communities have limited access to jobs, services, and transportation options. Long, expensive commutes to distant employment centers can eat up a large percentage of the family budget, or families have to live sparsely on the small amount of local work available. People who don't have access to personal vehicles or who do not drive, such as low-income residents and senior citizens, lack mobility and have even less access to jobs, healthcare, and other services.

---

[1] U.S. Department of Agriculture Economic Research Service. Rural Population and Migration Briefing Room. http://www.ers.usda.gov/Briefing/Population/.

[2] International City/County Management Association. *Putting Smart Growth to Work in Rural Communities.* http://www.icma.org/ruralsmartgrowth.

Other rural communities located close to metropolitan areas or amenities such as ski areas, national parks, and other tourist destinations are struggling to preserve their rural character in the face of growth pressures. These places are experiencing the conversion of farmland and natural land to development, which has an impact not only on the environment, but also on resource- and tourism-based economies. The new property development in these communities is often spread out, resulting in increasing demands for infrastructure in places where it is difficult and costly to provide.

Rural communities often lack the capacity or financial resources to address these issues. Some small, rural jurisdictions have limited local government staff, experience, or funding, which can mean few resources dedicated to comprehensive planning, regional collaboration, and other efforts to identify shared community goals and visions that can help shape growth and development. What's more, rural communities may lack access to private and public capital, making it difficult for them to obtain funds for economic development and revitalization. For example, philanthropic organizations that exist in larger communities are less present in rural areas, reducing resources that might assist local governments and organizations. The result can be development that fails to take advantage of the communities' assets, has limited long-term benefits, and creates long-term costs for the community.

An increasing number of rural communities are looking for development approaches beyond the conventional dispersed land use patterns that make it difficult for them to meet their fiscal, social, public health, and environmental goals. They are using a range of strategies to pursue economic opportunities while maintaining the rural character that residents value.

Sustainable communities approaches are as diverse as rural America itself. Communities select the most appropriate approaches for their context and adapt them to respond to local needs and interests. Some places are exploring new ways to generate income from working lands with the development of renewable energy facilities, including wind farms and solar panels. Others are directing public and private investments to main streets and village centers. Still others are planning and building walkable, convenient, and affordable neighborhoods. As the case studies in this report show, rural communities are finding solutions that allow them to take advantage of their assets, attract and retain businesses and residents of all ages, and ensure that economic development results in lasting improvements.

How the Livability Principles Support Rural Communities

Rural America is tremendously diverse—economically, demographically, and environmentally—and approaches to supporting sustainable rural communities should be equally varied. The Livability Principles guiding the Partnership for Sustainable Communities provide a useful policy framework for investments to support sustainable rural communities. The six principles are listed below, along with descriptions of how they can enhance economic prosperity and quality of life in rural places.

Livability Principles

Provide more transportation choices.

Many rural communities have limited transportation options. Rural roadways are often not designed to accommodate multiple modes of transportation, particularly walking, bicycling, and transit. Rural areas have also seen intercity bus service reduced over the past decades. This lack of options can limit access to jobs, medical care, and educational opportunities. In particular, seniors, low-income, and disabled persons living in rural America may be unable to reach necessary resources. For those who do drive, commutes to distant employment centers can be time consuming and require a large percentage of the family budget to be spent on transportation.

Residents of rural communities, like their counterparts in urban and suburban areas, benefit from neighborhoods that foster healthy and convenient walking, bicycling, and public transportation where feasible. Many rural communities were built on transportation corridors such as state highways, rail lines, or rivers and traditionally had compact, mixed-use designs with interconnected street networks that made it easy to walk or bicycle between neighborhoods and downtown. Village centers were ideal locations for regional transit services to pick up passengers. This foundation for expanded transportation choice both within and between towns still exists in many places. Looking at rural transportation through an intra- and inter- community lens can help guide investments from HUD, DOT, EPA, USDA, and other federal agencies.

In addition, intercity and regional mobility are drivers of economic growth in rural communities. Well planned transportation systems improve the quality of life and economic attractiveness of small towns by providing access to regional job markets, facilitating the transport of locally made goods to markets, and bringing tourists and other consumers to community businesses.

Promote equitable, affordable housing.

Some rural communities lack housing options. Much of their housing stock may be aging, resulting in low energy efficiency and high utility costs. Communities that offer a variety of housing types, such as single-family homes, townhouses, duplexes, and apartments in varying price ranges, are best positioned to attract and retain residents at all life phases—from single-person households to young families to retirees. The location of new housing can also provide a competitive advantage, as homes that are near schools, jobs, shopping, and services reduce residents' combined housing and transportation costs. Housing integrated into commercial areas, such as residences above first-floor stores and offices on main streets, may make it more convenient and affordable for residents to reach daily destinations while providing a local consumer base for businesses.

Enhance economic competitiveness.

Rural communities and small towns can thrive only if there are employment opportunities that support a good standard of living. While rural incomes may be substantially lower than those in metropolitan areas, rural regions possess unique resources and opportunities for economic development. Farms, ranches, renewable energy production facilities, and recreational amenities such as national parks and national forests all have economic value for rural communities. Innovations in agriculture can expand local and regional markets for agricultural products, resulting in more diverse, resilient economies. Continued expansion of broadband can also help strengthen and diversify rural economies, opening up new markets, connecting residents to job centers in larger communities, and reducing the need to travel to conduct business.

Federal investments are most effective if they are made in accordance with a community's economic vision. As a result, the Partnership should support rural communities' efforts to identify their competitive advantages through planning and visioning efforts.

Support existing communities.

Rural American communities are largely defined by their relationship to the agricultural and natural landscape, so conserving working and natural lands is a key strategy for protecting quality of life and the long-term economic viability of farming, forestry, tourism, and other natural resource-based activities. Redevelopment in small towns should support economic vitality without sacrificing the beauty and utility of the surrounding landscape. Rural America is home to many once-vibrant main streets with historic buildings and vacant commercial properties. Channeling investments into existing main streets can revitalize infrastructure and spur new economic opportunities. These assets can be the foundation for place-based economic development that promotes rural wealth creation. Outside of towns, directing and prioritizing investments can also meet community goals. For example, improving water and wastewater systems can protect subsurface fresh water sources and help farm families ensure the viability of agricultural operations.

Coordinate and leverage federal policies and investment.

Given the size and scale of rural communities, federal investments can make a significant impact, so it is critical that they support community goals and are coordinated across agencies. Federal investments are catalysts for additional public and private investment and can reinforce—or counteract—community plans that help guide development. However, many rural communities lack the capacity or resources to create plans and policies that codify their goals. As a result, the Partnership can support rural communities' efforts to craft visions for future development and to create and implement plans and policies that guide public and private investments.

Coordinating housing, transportation and environmental policies and funding produces better results for local residents. It uses taxpayer money more efficiently, meeting multiple economic, environmental, and community objectives with each dollar spent. To make their programs work as well as possible for the nation's communities, federal agencies must remove barriers to collaboration and provide opportunities to leverage funding.

For instance, federal investments in renewable energy development can improve economic conditions in rural America by creating jobs and reducing carbon emissions and dependence on oil imports. Where companies are investing in renewable energy facilities, coordinating transportation and infrastructure funding can ensure that employees have the housing and amenities they need, communities offer a high quality of life to attract a strong workforce, the environment is protected, and economic advances are sustainable. This same integrated approach can create opportunities in other areas such as recreation, tourism, and local and regional food systems.

Value communities and neighborhoods.

Rural communities and small towns should be valued for their distinctive and historic features. Communities that conserve and build upon these resources, such as historic downtowns and main streets, important natural features, and long-standing cultural and religious institutions will be better positioned to enhance quality of life for their residents. Iconic rural landscapes are often defined by farmsteads, historic barns, and working agricultural structures—visual representations of American agricultural traditions. Historic preservation, adapting old structures for new purposes, and designing to complement local character will strengthen existing communities while contributing to renewed economic vitality.

HUD, DOT, EPA, and USDA Programs at Work in Rural Communities

Working with its partners, the federal government has for many years played a significant role in rural America. The Partnership agencies and USDA have a variety of programs that support economic development, infrastructure, and conservation in rural communities. While some of these programs reflect the Partnership's objectives of enhancing economic competitiveness, supporting existing communities, and coordinating federal activities, others can be implemented without regard for the location of investments, their impacts on economic development and environmental outcomes, or potential leveraging of related federal, state, and local efforts. The effectiveness of these programs could be increased by better coordinating them with other federal efforts and incorporating the Livability Principles where appropriate.

This section lists some of the programs affecting the natural and built environments and economic development in rural communities. Each agency also has many other programs that can be helpful to rural communities. For more information, see each agency's website.

U. S. Department of Agriculture

USDA is comprised of multiple agencies that support agriculture, natural resources, and rural development. USDA's current strategic plan reflects a commitment to rural sustainability:

> *USDA is working to enhance the livability of rural communities. The department uses 21st-century technology to rebuild infrastructure, ensure that rural residents have decent housing and homeownership opportunities, clean water, adequate systems for handling waste, reliable electricity and renewable energy systems, and critical community facilities including health-care centers, schools, and public safety departments. USDA also helps communities invest in strategic green infrastructure planning and protection of critical natural resources.*[3]

While much of the department's work affects rural communities, three agencies stand out:

- Rural Development: Through its Rural Housing Service, the Rural Business-Cooperative Service and the Rural Utilities Service, Rural Development provides loans, loan guarantees, grants, and technical assistance to rural America. Resources are available for housing, community facilities, renewable energy and energy conservation, utility services (energy, water, and waste water), broadband, economic development, and capacity building. Rural Development administers over $20 billion per year in loans, loan guarantees and grants.

- Natural Resources Conservation Service (NRCS): Recognizing that over 70 percent of the United States' land area is privately owned, the Natural Resources Conservation Service partners with landowners to support and implement conservation planning that will result in productive lands and healthy ecosystems. NRCS brings expertise in watershed management and soil science, and its programs can help rural communities manage diverse landscapes.

---

[3] U.S. Department of Agriculture. *Strategic Plan, Fiscal Years 2010-2015.* http://www.ocfo.usda.gov/usdasp/sp2010/sp2010.pdf.

- Forest Service: The mission of the Forest Service is "to sustain the health, diversity, and productivity of the Nation's forests and grasslands to meet the needs of present and future generations."[4] National forests comprise 193 million acres across the country[5] and are an important presence in many rural communities, responding to many community needs from recreational opportunities to mining and timber resources to regional economic development. In areas where extensive public ownership of national forests affects the local tax base, the Forest Service provides direct financial support for community needs such as schools.

USDA is also spearheading various regional economic development initiatives. For instance, the Stronger Economies Together program aims to strengthen the capacity of rural regions to collaboratively develop and implement economic development blueprints. In another example, the first stage of the Great Regions initiative provided Rural Business Opportunity Grants to economic development projects that could have regional impacts, with the goal of further supporting these efforts through other USDA programs.

U.S. Department of Housing and Urban Development

Many HUD programs operate in rural as well as metropolitan areas, but their funds might pass through state agencies or other entities to rural communities, making it less evident that the funding comes from HUD. HUD invests around $6.2 billion per year in rural areas. The majority of these funds provide affordable housing to low-income residents. More than 800,000 families in rural communities currently receive assistance through HUD's Housing Choice Voucher, Public Housing, and Federal Housing Administration (FHA) Multifamily programs, with assistance totaling more than $4 billion per year.

The State and Small Cities Community Development Block Grant (CDBG) program enables rural communities to obtain grant funds for infrastructure, equitable affordable housing, economic development, and community planning. The State CDBG program provides about $840 million per year for economic development and other public investments in rural areas through state governments. The largest investments through the State and Small Cities CDBG are in public infrastructure, particularly for water and sewer projects, keeping Main Streets across rural America viable and directly supporting over 8,500 jobs per year. CDBG funds can also be used as the local match for other federal funding. HUD also funds over $500 million annually in affordable housing and homeownership programs in rural areas through another block grant to states, the HOME Investment Partnership. These two block grant programs also provide crucial additional funds to support communities rebuilding from disasters, serving as the vehicle for the supplemental appropriations Congress makes available to presidentially declared disaster areas.

HUD's Rural Housing Stability Grant Program assists individuals and families who are homeless, in danger of losing their homes, or in the worst housing situations in the geographic area – an increasing problem in rural communities. HUD's FHA insures more than $220 billion in mortgages,

---

[4] U.S. Forest Service. About Us – Mission. http://www.fs.fed.us/aboutus/mission.shtml.
[5] U.S. Forest Service. http://www.fs.fed.us/.

allowing over 1.5 million first-time homebuyers and other qualified families in rural areas to purchase their own homes. The FHA similarly insures loans on nursing homes, assisted living facilities, board and care facilities, and acute care hospitals, including a current total of over $545 million in loans to hospitals designated "critical access hospitals" in rural areas. Additionally, HUD ensures the quality of over 40,000 manufactured homes a year, more than 60 percent of which are located in rural areas. Less waste in construction and better energy efficiency make these homes more affordable to buy and live in.

HUD's Indian and Native Hawaiian Housing Block Grants, Home Loan Guarantees, and Community Development Block Grants support economic development and almost 40,000 homes, the single largest source of funding for housing on Indian tribal lands today at over $780 million per year. HUD supports these communities and their right to self-determination by allowing the recipients to design and implement housing programs according to local needs and customs.

HUD also administers several programs that link housing and economic development in rural areas. HUD's Rural Housing and Economic Development Program is designed to address problems of poverty, inadequate housing, and lack of economic opportunity in rural communities that are outside of metropolitan regions and have populations of 20,000 or less. It specifically focuses on high-need communities such as those in the Lower Mississippi Delta Region, the Colonias, and Appalachia, as well as federally recognized Indian tribes and seasonal farmworker communities. The program develops state and local capacity to support innovative rural housing and economic development approaches. Grants are awarded directly to local rural nonprofits, community development corporations, federally recognized Indian tribes, state housing finance agencies, and state community or economic development agencies.

Since fiscal year 1999, HUD's Office of Rural Housing and Economic Development has received an average of $22 million annually for the Rural Housing and Economic Development Program. Appropriations increased from $27 million in fiscal year 1999 to $248.7 million in fiscal year 2009, and grantees leveraged $850.9 million over that period. Grant awards increased from 91 in fiscal year 1999 to 964 in fiscal year 2009. Rural Housing and Economic Development grantees have created 13,005 jobs, trained 38,347 people, created 2,058 new businesses, assisted 5,557 existing businesses, constructed 8,595 housing units, and rehabilitated 9,267 housing units.

On a smaller scale, Rural Innovation Fund grants build homes and community facilities and support job-training programs and other economic development projects. This funding to federally recognized Indian tribes, local rural non-profits, community development corporations, and state housing and economic development agencies lets communities address local issues and builds their capacity to serve residents.

Launched in 2004, HUD's Rural Gateway serves as a clearinghouse of innovative ideas on rural housing, economic development, and revitalization, with a specific focus on communities in the Colonias, the Lower Mississippi Delta Region, Appalachia, and federally recognized Indian Tribes, as well as seasonal farmworker communities. The Rural Gateway builds the capacity of local, state, and regional organizations working on housing, economic development, and infrastructure development in rural areas. It also serves as a promoter and facilitator of private sector based partnerships to support housing, economic development, infrastructure, and capacity building activities.

HUD has historically been a key federal resource for community planning, and the Sustainable Communities Regional Planning Grant and Community Challenge Planning Grant programs, recent Partnership initiatives, focus on helping communities create plans that integrate economic development, housing, and transportation. These grant programs provide a useful example of broadly applicable initiatives that set aside a portion of funding for rural and small communities. For instance, $28 million in Sustainable Communities Regional Planning Grants was awarded to regions with populations less than 500,000 and $15 million in Community Challenge Planning Grants went to rural places with fewer than 200,000 people. The interest from rural areas in the first year was great—52 percent of applicants to the Sustainable Communities Regional Planning Grant Program were from small towns and rural areas.

U.S. Department of Transportation

Like HUD, most programs in DOT, including safety research, highway and transit construction, and transportation planning, have a rural component. Most federal transportation funding is distributed by formula to state and local transportation agencies.

The Federal Transit Administration (FTA) administers a variety of programs that provide access to public transportation in rural communities, particularly to prevent the economic and social isolation of elderly and low-income residents and to support intercity mobility. FTA's formula grant programs provide over $500 million for public transportation in and between rural communities, in addition to targeted technical assistance for rural transit providers, grants to tribes for transit and roads, and grants that support sustainable transportation for visitors to national parks and federal lands. Since 1979, FTA has provided grants to states under the Section 5311 Non-urbanized Transit Program to establish and maintain transit systems in communities with populations under 50,000.

The Rural Transit Assistance Program provides funding for training and technical assistance projects and other support services tailored to meet the needs of transit operators in rural areas. Additionally, the Tribal Transit Program provides approximately $45 million in direct funding to federally recognized tribes to support tribal public transportation in rural areas. The tribes can use this funding for capital, operating, planning, and administrative expenses for transit projects that meet the growing needs of rural tribal communities.

The Paul S. Sarbanes Transit in Parks Program provides funding for sustainable transportation systems, such as shuttle buses, rail connections, and bicycle trails in America's national parks, national forests, and wildlife refuges. The program seeks to conserve natural, historical, and cultural resources; reduce congestion and pollution; improve visitor mobility and accessibility; enhance the visitor experience; and ensure access to everyone, including persons with disabilities. FTA awarded $27 million through this program in 2010.

The Federal Highway Administration (FHWA) plays an important role in supporting the transportation needs of rural communities with its investments in America's highway and road network, including state highways that often serve as main streets through rural towns and villages.

In fiscal year 2008, nearly 39 percent of all federal highway funds obligated—approximately $13.7 billion—were for highways classified as rural.

FHWA also administers the Highway Safety Improvement Program (HSIP) to reduce traffic fatalities and injuries on public roads. The High Risk Rural Roads Program was established through a set-aside from each state's apportionment of HSIP funds for construction and operational improvements on high-risk rural roads. A total of $90 million is set aside nationally per year and is applied proportionally from the states' HSIP apportionments.[6]

The Federal Lands Highway Program provides transportation infrastructure for the 30 percent of America that is federally owned (national parks, forests, monuments, reserves, rangelands, etc.). Similarly, the National Scenic Byways Program invests in transportation corridors to support the environmental and cultural features that make them economically valuable.

With a more regional focus, FHWA also supports the Appalachian Development Highway System Program. This program provides funds for the construction of highways in 13 states in Appalachia to promote economic development and meet the region's transportation needs. The Appalachian Development Highway System is 76 percent complete.

Additionally, FHWA supports freight movement, which often passes through rural areas and is critical for bringing rural products to market. Freight operations and planning are dealt with across many of FHWA's program areas. The agency also sponsors peer-to-peer exchanges and seminars on freight movement and planning, and on linking freight transportation and livability.

The FTA and FHWA jointly administer the State Planning and Research Programs, through which states are required to conduct comprehensive and collaborative intermodal statewide transportation planning that facilitates the efficient movement of people and goods. As part of this process, states are required to consult with officials from places outside of metropolitan areas.[7]

The Federal Railroad Administration administers the Railroad Rehabilitation & Improvement Financing Program which provides direct federal loans and loan guarantees to finance development of railroad infrastructure that can particularly benefit large projects in rural areas.

Rural areas also compete successfully in DOT's discretionary livability grant programs. In 2009 and 2010, with the support of the Partnership, the TIGER (Transportation Investments Generating Economic Recovery) and TIGER II programs provided approximately $288 million in funding to support the planning and construction of transportation infrastructure in rural areas.

---

[6] U.S. Department of Transportation Federal Highway Administration. What We Do. http://www.fhwa.dot.gov/whatwedo/topics/.

[7] U.S. Department of Transportation Federal Highway Administration. What We Do. http://www.fhwa.dot.gov/whatwedo/topics/.

U.S. Environmental Protection Agency

EPA's mission is to protect the environment and public health, and a healthy environment is essential to a healthy economy. Programs that most explicitly integrate environmental, economic, and community outcomes in rural places include technical assistance offered by the Office of Sustainable Communities; support for wastewater and drinking water infrastructure provided by the Office of Water; and brownfields assessment, cleanup, and area-wide planning grants from the Office of Brownfields and Land Revitalization. These programs do not formally set aside funds for rural communities, but they have worked with many rural places.

For example, the Office of Sustainable Communities has provided intensive technical assistance to rural communities through its Smart Growth Implementation Assistance program as well as offering more targeted, short-term help on growth and development challenges in small towns through its Sustainable Communities Building Blocks program. Fifteen of the 31 Building Blocks technical assistance projects served small towns and rural communities, bringing approximately $180,000 in technical assistance services. The Office of Sustainable Communities also provides funding for the Governors' Institute on Community Design, which offers technical assistance to governors and their staffs, some of whom have asked for help with rural issues. The office has funded and helped to create the Smart Growth Network publication *Putting Smart Growth to Work in Rural Communities*,[8] which highlights smart growth strategies that leaders from rural communities and small towns can use to help guide growth while protecting natural and working lands and preserving rural character. Smart growth is development that is good for the economy, the environment, and the community, providing more choices for residents, greater opportunity across the community, good return on public investment, and clean air and water.[9] Smart growth techniques look different in different places because they are meant to be adapted to local needs, but many communities use smart growth strategies to foster neighborhoods that have stores, offices, schools, and houses of worship near homes; to preserve open space for agriculture, recreation, and aesthetic value; and to ensure that people can find a safe, convenient, and affordable place to live.

EPA's Brownfields Program, in the Office of Brownfields and Land Revitalization, empowers states, communities, and other stakeholders in economic redevelopment to work together to prevent, assess, clean up, and reuse brownfields. A brownfield is a property, the expansion, redevelopment, or reuse of which may be complicated by the presence or potential presence of a hazardous substance, pollutant, or contaminant. Brownfields can exist in urban and rural communities. The Brownfields Program offers grants to support revitalization efforts by funding environmental assessment, cleanup, and job training activities. Brownfields Assessment Grants provide funding for brownfield inventories, planning, environmental assessments, and community outreach. Brownfields Revolving Loan Fund Grants provide funding to capitalize loans that are used to clean

---

[8] International City/County Management Association. *Putting Smart Growth to Work in Rural Communities.* http://www.icma.org/ruralsmartgrowth.

[9] Smart growth is further described by the ten smart growth principles, developed by the Smart Growth Network based on the experiences of urban, suburban, and rural communities around the nation that have used smart growth approaches to create and maintain great neighborhoods. See the Smart Growth Network website for a discussion of these principles: http://www.smartgrowth.org.

up brownfields. Brownfields Job Training Grants provide environmental training for residents of brownfields communities. Brownfields Cleanup Grants provide direct funding for cleanup activities at certain properties with planned green space, recreational, or other nonprofit uses. In 2010, the Brownfields Program worked with the Partnership to give out Area-Wide Planning Grants to help selected communities create a shared vision for brownfields redevelopment that will inform cleanup decisions. Five out of 23 of those grants, representing nearly $1 million in funding, supported rural communities with populations less than 20,000.

The Office of Water provides grants to states to operate revolving loan programs that provide low-interest financing for wastewater, drinking water, and other water quality projects. In 2010, these programs issued guidance recommending that states make funding decisions that are consistent with the Partnership's Livability Principles and discourage expanding infrastructure to accommodate growth if there are available facilities in existing communities. In fiscal year 2009, 78 percent of the Clean Water State Revolving Fund assistance agreements—approximately $1.2 billion—were established with communities of fewer than 10,000 people.[10] About $608 million in assistance was provided to communities of fewer than 10,000 people through the Drinking Water State Revolving Fund in fiscal year 2009.[11]

Other EPA programs at work in rural areas include AgSTAR, which advances the capture and use of biogas at livestock facilities; Smartway Transport, which works with the freight sector to improve energy efficiency, reduce greenhouse gas and air pollutant emissions, and improve energy security; and the U.S.-Mexico Border Water Infrastructure Grant Program, which provides grants for the planning, design, and construction of wastewater and drinking water facilities to communities within 60 miles of the border.

---

[10] U.S. Environmental Protection Agency. *Clean Water State Revolving Fund Programs 2009 Annual Report.* http://water.epa.gov/grants_funding/cwsrf/upload/2009_CWSRF_AR.pdf.
[11] U.S. Environmental Protection Agency. *Drinking Water State Revolving Fund: 2009 Annual Report.* http://water.epa.gov/grants_funding/dwsrf/upload/dwsrf-annualreport2009nov2010.pdf.

Performance Measures for Success

Communities of all sizes are using performance measurement to better understand the impacts of their decisions. Performance measures document changes in human behavior, demographics, economic trends, or development patterns. By translating data and statistics into a succinct and consistent format, performance measures quantify the degree to which programs, policies, and investments achieve community goals. Performance measures allow decision-makers to quickly observe the expected effects of a proposed plan or project or to monitor trends in its performance over time.

Along with helping rural communities track progress toward their own sustainable communities goals, performance measurement can also aid the Partnership agencies in assessing the effectiveness of their investments in rural communities and small towns. Performance measures can help HUD, DOT, EPA, and USDA translate the Livability Principles into concrete outcomes, target their resources toward planning and capital programs that support sustainable communities, and evaluate federal initiatives.

Rural communities, given their distinctive characteristics, require customized performance measures. The measures provided here are suggestions for communities or regions interested in performance evaluation at the community or regional scale. Each measure can also be used or adapted to assess the performance of federal programs at a national scale. However, some measures may require local data that are not available in a nationally consistent format. For example, while the U.S. Census counts housing units every 10 years, more timely and geographically precise information on the location of new home construction would have to be acquired from each county building permitting office or tax assessor. As a result, it is difficult to accurately measure at a national scale the percentage of new housing units built on previously developed land or near rural town centers, or the average density of new residential development. Likewise there is no single consistent database describing land use or land value at a parcel level. As a consequence, it is impossible to assess how different places across the country arrange and balance residential, commercial, industrial, and agricultural uses, or how property values change over time.

The following framework for performance measurement is organized in terms of broad goals and specific strategies that can help attain each objective. Each rural community can choose a different set of strategies that best fits its opportunities and challenges. The implementation measures evaluate the effectiveness with which each strategy is pursued. Other indicators can be used to track a community's progress toward the broader goals. These measures reflect changes in behavior or outcomes on the ground that would be anticipated to result if strategies are implemented successfully. All the measures described here are examples that can be helpful to rural communities trying to become more environmentally and economically sustainable, but they are a starting point, not a definitive list.

Goal 1: Promote Rural Prosperity

Create an economic climate that enhances the viability of working lands, preserves natural resources, and increases economic opportunities for all residents in rural communities.

Strategies	Implementation measures[12]	Other indicators
Pursue regional collaboration	• Creation of a regional economic development plan based on a clear understanding of comparative economic advantages and existing or emerging economic clusters • Implementation of a regional economic development plan	• Integration of regional economic development plan with transportation, housing, land use, natural resources, workforce development, and other regional or local plans
Cultivate economic development that promotes the sustained economic potential of working rural lands	• Successful development of supplementary economic uses for rural lands and their byproducts (e.g., wind farms, biomass power generation) • Implementation of policies to promote sustained economic viability of agricultural and natural resource land uses	• Rate of agricultural and natural resource land lost to development • Percentage of prime rural land lost to development • Percentage of prime agricultural land placed under permanent conservation easement
Cultivate economic development that sustains a high quality of life in rural communities	• Creation of economic development plans or strategies that are based on unique assets and include measurable goals • Implementation of policies to promote natural resource conservation and environmental quality	• Percentage of jobs at region's three largest employers[13] • Percentage of jobs in small- to medium-sized firms • Percentage of jobs in locally owned firms • Percentage of new jobs in high-wage occupations • Regional exports • Growth of sectors that are part of asset-based or cluster development

---

[12] These implementation measures (unlike those in Goals 2, 3, and 4) focus not on outcomes on the ground, but rather the development and implementation of plans or policies that can shape those outcomes. At the community scale, they are measured nominally (e.g., whether a plan/policy is in place). They can also be adapted for national-scale program evaluation. One example of a national measure might be the percentage of grant-receiving communities that have created a regional economic development plan that is based on a clear understanding of comparative economic advantages and existing or emerging economic clusters.

[13] Sustaining long-term economic opportunity in rural communities sometimes means increasing the number of employers. Rural communities that rely upon a few major employers are less economically resilient when one of those employers chooses to reduce or close down operations. Therefore an economic development strategy might encourage increasing the percentage of jobs in small to medium-sized firms or locally owned firms that are more likely to have a long-term interest in the community.

Goal 2: Support Vibrant and Thriving Rural Communities

Enhance the distinctive characteristics of rural communities by investing in rural town centers, Main Streets, and existing infrastructure to create places that are vibrant, healthy, safe, and walkable.

Strategies	Implementation measures	Other indicators
Invest public funds in existing rural communities	• Percentage of public investments[14] in rural areas spent on projects within ½ mile of rural town centers[15]	• Common elements in transportation, land use, housing, economic development, natural resources, and water plans promoting public investment in existing communities
Encourage private-sector investment in existing rural communities	• Percentage of new or rehabilitated housing units within ½ mile of rural town centers • Percentage of new commercial development (or major rehab) within ½ mile of rural town centers • Percentage of new housing units built on previously developed land • Percentage of new commercial development on previously developed land	• Percentage of households within ½ mile of rural town centers • Percentage of employment within ½ mile of rural town centers • Number of jobs within ½ mile of rural town centers • Number of brownfields remediated for redevelopment
Make it easy to build compact, walkable, mixed-use places	• Percentage of new homes built in mixed-use neighborhoods • Average density (units per acre) of new residential development	• Percentage of households with walkable/convenient access to stores, services, parks, and/or schools[16] • Street network connectivity of new development (block length or number of three- or four-way intersections)

---

[14] This measure could be adapted to evaluate either investments from a single state or federal program or a collection of different programs.

[15] The term "rural town center" can refer to historic Main Streets as well as newer developments in which a variety of jobs, housing, retail, and services are concentrated. One potential way to identify rural town centers is to use Census-designated urban area boundaries for towns or cities of between 2,500 (the minimum) and 49,999 in population. Additionally, the Partnership for Sustainable Communities working group on Performance Measurement is developing a national dataset to define the locations of activity centers in both urban and rural communities across the U.S.

[16] The range of services available in a rural community will depend on that community's population. For example, a community of 15,000 residents might be capable of supporting a full-service grocery store while a community of 1,000 residents might not be. Therefore, this indicator should be adjusted to reflect realistic expectations and local context.

Goal 3: Expand Transportation Choices

Create communities where everyone—including elderly, disabled, and low-income residents—can conveniently, affordably, and safely access local and regional goods and services.

Strategies	Implementation measures	Other indicators
Increase multimodal mobility and access for rural communities	• Percentage of non-urbanized area population covered by demand-response service[17] at least three days per week[18] • Availability of fixed route transit service in key travel corridors, where appropriate[19] • Availability of scheduled intercity bus or rail service	• Average number of daily scheduled intercity bus and rail departures from a rural town center to larger communities where health care, schools, jobs centers and other regional services are available • Transit trips per capita
Design roadways that support all modes of travel: transit, biking, walking, and automobile	• Percentage of new or improved roadways (by mile) that include sidewalks and/or bicycle/pedestrian infrastructure • Adoption of "complete streets"[20] policy in the long-range/short-range transportation plan	• Biking mode share for trips to work • Walking mode share for trips to work • Transit mode share for trips to work • Pedestrian and bicyclist fatality rate[21]

---

[17] Demand-response service is a form of public transportation with small or medium-sized vehicles operating on flexible routes and schedules according to passenger needs. An example is Dial-a-Ride service.

[18] Note that this measure should be adapted as appropriate for the size of the community. Smaller rural communities may only offer demand-response service such as paratransit while larger rural communities may be able to support and benefit from fixed-route transit service along key corridors.

[19] See previous note.

[20] "Complete streets" are roadways designed and operated to enable safe and comfortable access and travel for all users.

[21] Note that this indicator alone is a poor measure of the success of programs seeking to promote walking and biking in rural communities. For example, a community where no residents walk or bike will have a low fatality rate. Nevertheless, in communities with limited data regarding bike and pedestrian activity, this measure may provide one useful perspective on progress towards improved walking and biking conditions.

Goal 4: Expand Affordable Housing

Create communities where everyone—including elderly, disabled, and low-income residents—can afford housing and transportation expenses.

Strategies	Implementation measures	Other indicators
Increase in affordable housing near rural town and employment centers	• Number of affordable for-purchase and rental homes in or near rural town centers • Implementation of policies to ensure that housing is affordable to working families, the elderly, and low-income residents	• Percentage of low-income households within a 30-minute commute of regional employment centers • Median household housing plus transportation costs

Conclusion and Next Steps for the Partnership

The Partnership for Sustainable Communities will continue working to ensure that its policies, programs, and investments support rural communities that are economically resilient, provide good quality of life for residents, and have healthy environments. To strengthen federal support of rural communities and use Partnership resources efficiently, HUD, DOT, EPA, and USDA will consider the following next steps.

Short Term (3-6 months)

- Form a Rural Implementation Group made up of staff from the four agencies to implement the steps identified in this report.
- Create a *Leveraging the Partnership for Rural Communities* guide, modeled on the *Leveraging the Partnership* document,[22] describing funding and technical assistance programs available to rural communities from each of the four agencies.
- Facilitate collaboration between Rural Development state staff and Partnership regional teams on ongoing HUD, DOT, EPA, and USDA projects with rural components.
- Determine the potential for philanthropic resources to help build rural capacity for strategic planning and implementation of specific projects.
- Explore the feasibility of a capacity-building workshop on sustainable communities strategies for rural grantees of HUD's Regional Planning Grant program.
- Conduct outreach to rural stakeholders on the Partnership's activities.

Medium Term (6-12 months)

- Post online case studies on rural communities that have used smart growth and sustainable communities approaches to achieve job growth, resource protection, and housing and community facility improvements.
- As follow-up to the Rural Roundtable held by the Rural Work Group in August 2010, conduct additional listening sessions with agency leadership and stakeholder groups.
- Collaborate with USDA Economic Research Service to identify how its online Rural Atlas can become an effective information tool for current and future Partnership grantees.[23]
- Engage in collaborative technical assistance and/or grant implementation.
- Consider ways to streamline and improve grant processes so that rural communities can access federal resources more easily.
- Address capacity issues related to grant writing and planning capacity in rural communities.
- Identify regional collaboration opportunities that will help the four agencies assess how well their programs work together.
- Prepare guidelines for planning effective transit programs in rural areas.
- Support rural sessions and networking at the 2012 New Partners for Smart Growth conference.

---

[22] Partnership for Sustainable Communities. *Leveraging the Partnership.* http://www.epa.gov/smartgrowth/pdf/2010_0506_leveraging_partnership.pdf.
[23] U.S. Department of Agriculture Economic Research Service. Rural Atlas. http://www.ers.usda.gov/Data/RuralAtlas/index.htm.

Longer Term (12-18 months)

- Continue to incorporate the Livability Principles, as appropriate, into rural-focused community and economic development Notices of Funding Availability and grant applications.
- Evaluate the impacts of the Partnership's rural efforts through performance measurement.
- Deliver Partnership resources to rural communities through training and technical assistance.
- Explore how the Partnership can help support scenario planning workshops in rural areas.

Appendix A: Case Studies of Federal Support for Sustainable Rural Communities

Across the country, rural communities are creating development that strengthens their economies, takes advantage of assets like traditional Main Streets and agricultural lands, and provides residents with more housing and transportation choices. These case studies are examples of rural communities working with federal agencies to attain their quality of life, environmental, and economic goals. These case studies are in alphabetical order by state.

Federal Support for Sustainable Rural Communities
Case Studies

Grand Canyon National Park: Enhancing Visitor Experiences through Multimodal Transportation Improvements

Photos courtesy of the National Park Service

Location: Grand Canyon National Park, Arizona

Focus: Multimodal transportation improvements

Partners:
U.S. Department of Transportation
National Park Service

Funding:
Federal Lands Highway Program

Project Description:

To ensure positive experiences for the Grand Canyon National Park's 5 million visitors each year, the Federal Lands Highway Program supported enhanced shuttle services within and outside the park, bike rental facilities, pedestrian facility upgrades, and other transportation improvements.

Established in 1919, Grand Canyon National Park is an icon in the national parks system. The canyon itself includes over 277 river miles and can be as much as 18 miles wide and a mile deep. The park is home to breathtaking and unique geographic features and important archeological and cultural resources.

In 2007, the National Park Service and the Forest Service conducted the South Rim Visitor Transportation Plan Environmental Assessment to address the park's pressing traffic, parking, and access issues, specifically those in Grand Canyon Village, where many visitors stay. Most of the components of the plan have been or are being implemented.

Grand Canyon National Park: Enhancing Visitor Experiences through Multimodal Transportation Improvements

Project Components:

- A new shuttle route to transport visitors to the South Rim from the gateway community of Tusayan, seven miles outside the park.

- Expanded shuttle service from the visitor center to multiple South Rim destinations.

- Bike rental facilities at the Canyon View Information Plaza.

- Entrance station improvements to reduce long wait times entering the park.

- Improved shuttle stops, pedestrian improvements, roadway realignments, and new parking at the Canyon View Information Plaza.

- Intelligent Transportation Systems, which integrate communications and electronics technologies into transportation infrastructure to improve traveler information and enhance safety and mobility.

Livability Principles Addressed:

Provide more transportation choices: Recognizing the park's worsening traffic, parking, and visitor access, the plan's implementation increases the transportation choices for getting to and around the park. With enhanced shuttle service and bike rentals available, visitors have transportation options that allow them to connect more closely with their environment while reducing congestion. Additionally, the shuttle service provides another transportation option to park employees, many of whom live along its route inside the park.

Enhance economic competitiveness: Providing enhanced transportation services to nearby gateway communities can strengthen their economies. For example, the shuttle service increases visitors' access to the hotels and restaurants in Tusayan.

Support existing communities: The South Rim Visitor Transportation Plan works with neighboring communities such as Tusayan to provide transportation to and from the national park, improving residents' access to the park and visitors' access to the community.

Coordinate and leverage federal policies and investment: Grand Canyon National Park is coordinating with the Arizona Department of Transportation on a complementary streetscape improvement project in Tusayan.

For more information about this project, contact:

Elijah Henley, DOT, elijah.henley@dot.gov

Federal Support for Sustainable Rural Communities
Case Studies

USDA
Rural Development
Committed to the future of rural communities

Lake Village: Reusing a Historic Building to Support Downtown

Photos: Historic Lake Village. John Tushek Building before renovation. Courtesy of Aaron Ruby.

Location: Lake Village, Arkansas

Focus: Historic preservation and downtown revitalization

Funding:
USDA Community Facilities Program: $840,000
Arkansas Energy Efficiency Conservation Block
Grant Program: $750,000

Partners:
U.S. Department of Agriculture – Rural
Development Arkansas State Office
City of Lake Village

Project Description:

In 2010, the community of Lake Village, Arkansas, population 2,823, received funding to rehabilitate a historic structure in its town center in an effort to consolidate public service providers into one location and channel future development into the Main Street area of an economically distressed community.

Like many small communities whose main streets have declined, Lake Village had seen public and private investments migrate to the outskirts of town over the years, leaving Main Street a shadow of its once-vibrant self. In an effort to reverse that trend, Lake Village leaders explored ways to revitalize their community and decided that reusing an existing building, which is listed on the National Register of Historic Places, would be one way to provide a boost to the community. With the mayor, police, and court clerk all using inadequate spaces in separate buildings, the town hoped that combining those departments into one centrally located building would help provide services to the community more efficiently while also bringing people and economic activity back to Main Street.

Once complete, the historic John Tushek Building will be among the first LEED-certified buildings in Arkansas, will be the home of all the town's public service providers, and will be a gathering place that, in the coming years, can help attract other offices and businesses to locate on Main Street.

Lake Village: Reusing a Historic Building to Support Downtown

Community Outreach:

In the 1990s, the local chamber of commerce decided to retain a downtown location rather than relocate out of the town center to the nearby state highway. Supporting this decision, economic consultants recommended to the city council in 2006 that city administrative services be consolidated and located in the town center. The decision to redevelop the Tushek Building resulted from this strategic planning process.

Livability Principles Addressed:

Enhance economic competitiveness: By consolidating public services into one building, Lake Village will create a critical mass of employment downtown, which can help attract other businesses to Main Street and renew its vitality.

Support existing communities: Using an existing building is a more efficient use of scarce resources than building a new facility, and all the utilities needed to serve it, from scratch.

Coordinate and leverage federal policies and investment: Combining USDA-Rural Development and state funds is enabling the city to rehabilitate the Tushek Building using LEED development standards, which will reduce energy costs and advance the community's goal of revitalizing its Main Street.

Value communities and neighborhoods: Rehabilitation and reuse of the Tushek Building as a civic space is a testimony to the community's appreciation for this historic asset, as well as for their distinctive Main Street and the surrounding neighborhoods.

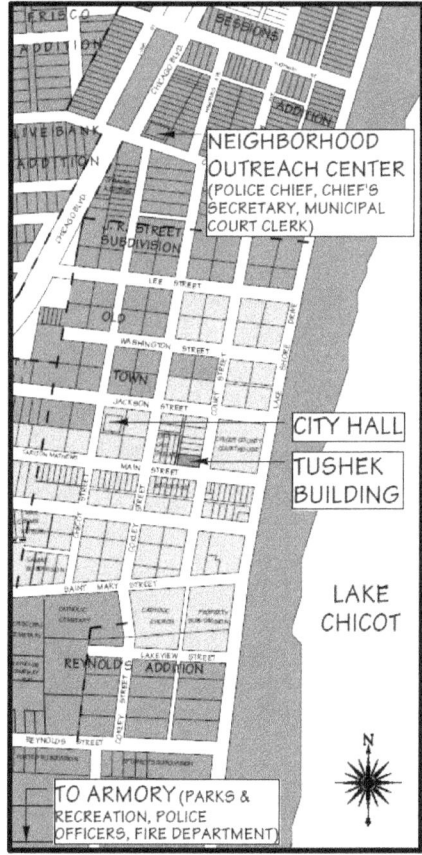

Images, top to bottom: USDA, City of Lake Village

For more information about this project, contact:
Steve Horsman, USDA-Rural Development Arkansas State Office, 870-367-8400, steve.horsman@ar.usda.gov

Federal Support for Sustainable Rural Communities
Case Studies

Waverly: Disaster Resiliency Through Smart Planning

Photos, left to right: Downtown Waverly Market Study Summary, Downtown Waverly Market Study Summary, EPA

Location: Waverly, Iowa

Focus: Disaster resiliency and mitigation through smart growth

Funding:
EPA: About $60,000
FEMA: About $5,000

Partners:
U.S. Environmental Protection Agency
U.S. Department of Agriculture – Rural Development
Federal Emergency Management Agency
Rebuild Iowa Office
Iowa Department of Economic Development
Iowa Northland Region Council of Governments
City of Waverly

Project Description:

The city of Waverly, Iowa (population 8,968) was one of several Iowa communities selected by EPA and FEMA to receive technical assistance to help recover from flooding that took place in June 2008. Recognizing an opportunity to prepare for future challenges, the city asked for assistance with conducting an audit of its policies and development regulations to assess whether the policies integrated smart growth concepts and approaches, identifying green infrastructure strategies that could connect vacant lots in the city as part of a larger open space plan, and exploring options for infill and affordable, mixed-income housing.

EPA, USDA, FEMA, and other partners assembled a technical assistance team of national experts in community design and planning. At the Waverly Smart Planning Workshop on May 26-27, 2010, the team worked with the community to help the city develop policies and project designs that could be incorporated into the city's comprehensive plan. The event included a tour of the city, meetings with stakeholders to discuss preliminary policy ideas, a community workshop to present draft policy ideas and project designs, and a community open house to gather feedback on refined policy ideas and project designs.

Based on the input gathered during the Waverly Smart Planning Workshop, the technical assistance team developed a memo outlining policy options and project design ideas that the city is now using to inform its Open Space Master Plan and the comprehensive plan, which is being revised. The city is already beginning to implement many of the concepts discussed at the workshop, including community gardens, complete streets with bicycle and pedestrian accommodations, mixed-use development, and affordable housing.

Waverly: Disaster Resiliency Through Smart Planning

Community Outreach:

Community involvement was an important part of the Waverly Smart Planning Workshop. The planning team conducted interviews with stakeholders prior to the workshop to help shape and refine initial options. The city of Waverly also conducted public outreach to ensure that stakeholders were represented throughout the workshop. The city's comprehensive plan revision process, which is building on the work that was done at the workshop, also involves significant public outreach.

Livability Principles Addressed:

Provide more transportation choices: The Smart Planning Workshop discussed strategies for better connecting Waverly's street grid, making it easier to walk and bicycle in the city.

Promote equitable, affordable housing: The workshop explored building affordable and workforce housing in areas that are adjacent to existing neighborhoods, allowing future residents to live close to jobs, schools, and other amenities.

Enhance economic competitiveness: The options developed in the workshop can position Waverly to attract and retain residents, to enhance local businesses, and to build on its existing economic assets.

Support existing communities: The policy options would support and enhance existing neighborhoods in the city.

Coordinate and leverage federal policies and investment: USDA Rural Development's involvement in the workshop provided an opportunity to explore how USDA funds could help implement the smart growth strategies discussed during the workshop.

Value communities and neighborhoods: The workshop highlighted many of Waverly's great qualities and provided options for the city to consider on how to build on those assets.

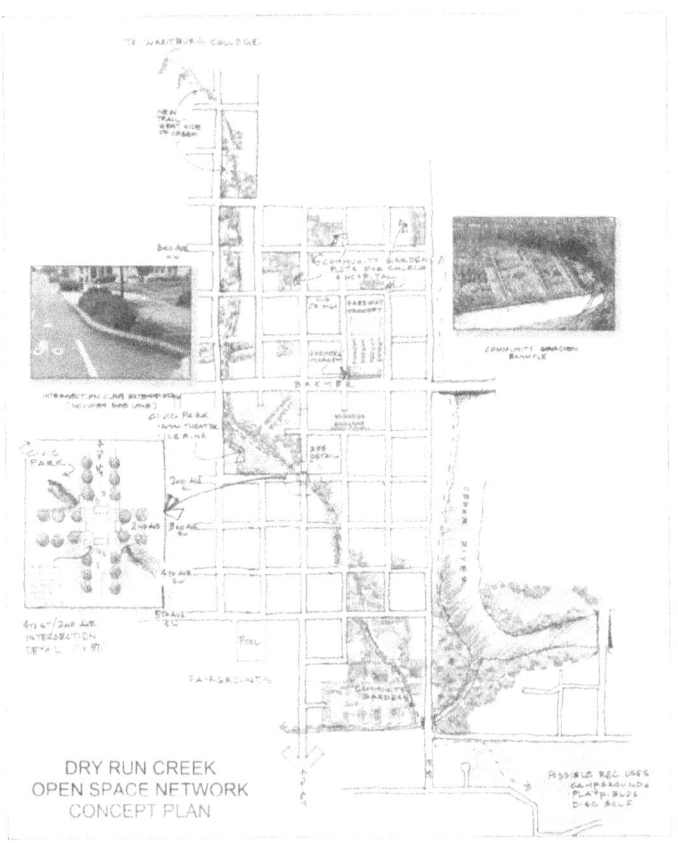

DRY RUN CREEK
OPEN SPACE NETWORK
CONCEPT PLAN

Photos. top to bottom: EPA. FEMA

For more information about this project, contact:
Stephanie Bertaina, EPA
202-566-0157
bertaina.stephanie@epa.gov

Federal Support for Sustainable Rural Communities
Case Studies

Greensburg: Rebuilding a Community with Green Design

Photos courtesy of EPA

Location: Greensburg, Kansas

Focus: Greensburg has embraced green building, sustainable design, and renewable energy strategies as the community rebuilds after a devastating tornado in 2007, including requiring public buildings to be certified LEED Platinum and developing a ten turbine wind farm.

Federal Partners:
U.S. Environmental Protection Agency
U.S. Department of Agriculture
U.S. Department of Commerce (Economic Development Administration)
U.S. Department of Energy (National Renewable Energy Laboratory)
Federal Emergency Management Agency
Small Business Administration

Project Description:

On May 4, 2007, Greensburg was hit by an EF-5 tornado which killed 11 people and destroyed 95% of the city. In the wake of the tragedy, those residents who chose to return to Greensburg decided to rebuild the city by embracing green building, sustainable design, and renewable energy. In December 2007, as part of its recovery plan, Greensburg passed a resolution requiring all new public buildings to achieve a LEED Platinum rating. These buildings utilize wind turbines, solar panels, high-efficiency windows, recycled materials, and other techniques which reduce energy consumption and save hundreds of thousands of dollars in energy bills. Additionally, the city receives its power from a ten turbine wind farm which provides enough energy to serve Greensburg and nearby communities.

Prior to the 2007 tornado, Greensburg had been facing economic challenges common to many other small Midwestern communities. Impacted by changes in the agricultural industry, by the year 2000 Greensburg was a struggling city of 1,500 people with a per capita income of around $18,000. After the destruction of the city, residents and local officials saw an opportunity to rebuild in a way that was "stronger, better, greener" – Greensburg's new motto.

Receiving assistance from many federal agencies as well as support at the state level, Greensburg has begun its path to a sustainable recovery. Following the new guidelines established for the city under the 2007 green building resolution, Greensburg has built its school, city hall, hospital, county building, courthouse, and an arts center. The city hall was built with bricks that were collected from a power plant that was destroyed by the tornado and also utilizes geothermal heating and cooling and solar panels for energy.

Greensburg: Rebuilding a Community with Green Design

The new K-12 school uses geothermal heat and a wind generator as well as other green systems that make the building 50 percent more efficient than if built under the traditional code. Additionally, in this time of fiscal challenges and constraints, Greensburg's new buildings provide a substantial amount of savings in energy costs. The new school saves an estimated $150,000 a year, the hospital around $120,000, and the courthouse $14,000. In addition to the city's new structures, Greensburg's Main Street has been redeveloped as a narrower, more walkable space. The street also utilizes a green stormwater design system that nourishes plants during the dry season with water collected and stored in underground cisterns. Main Street has also supported some of Greensburg's new businesses, including an insurance agency, coffee shop, home furnishings store, and others. The city established a "business incubator" to nurture new businesses by providing space at an affordable rent until the business is ready to expand.

Greensburg's wind farm, located three miles outside of the city, consists of ten 300 foot turbines. The wind farm produces 12.5 MW of energy, enough to generate power for the entire city as well as other nearby communities, making Greensburg a green energy provider for the region. Technical assistance for the wind farm, as well as for the city's master plan and energy-efficient buildings, was provided by the Department of Energy's National Renewable Energy Laboratory. FEMA was also instrumental in supporting the first phase of Greensburg's recovery efforts.

Following the disaster and the town's new commitment to sustainable redevelopment, the *Greensburg GreenTown* grassroots organization was established to support residents, businesses, and the local government in achieving its vision. GreenTown has launched a variety of programs, including technical assistance, educational trainings, fundraising initiatives for local sustainable development projects, and a GreenTour map and book for visitors and residents. GreenTown also created the "Chain of Eco-Homes" project, featuring model green homes built to educate local residents and attract visitors who pay to stay in the homes as overnight guests. Local residents have also adopted many of the green building techniques in their own new home construction, such as double-pane windows, thicker walls, solar panels, and geo-thermal heating.

Greensburg has embraced its role as a model green town and is turning its strategy into a commercial venture not only through its wind farm and green jobs, but also through a budding eco-tourism industry. While Greensburg has always had visitors to see its 109 foot deep well (the largest hand-dug well of its kind), the city's green buildings are becoming the community's newest attractions.

Photo: EPA

Visitors pay around $100 dollars per night to stay in the Chain of Eco-Homes. Tours are also given for visitors to learn about Greensburg's redevelopment plans.

Greensburg and GreenTown have worked hard to encourage other communities, especially those impacted by natural disasters, to adopt similar redevelopment strategies. Recently, a group of leaders from Reading, Kansas visited Greensburg to learn how to rebuild their town after a tornado hit in May 2011. Officials would like the city to serve as an example for other communities such as Joplin, Missouri and Tuscaloosa, Alabama seeking to rebuild sustainably after disasters.

For more information about this project, contact:
David Doyle, EPA
913-551-7667
doyle.david@epa.gov

Federal Support for Sustainable Rural Communities
Case Studies

Greening the Block in Bowling Green

Photos courtesy of HUD and the Housing Authority of Bowling Green

Location: Bowling Green, Kentucky

Focus: Home energy efficiency, access to opportunity

Funding:
HUD: About $1.28 million

Partners:
U.S. Department of Housing and Urban Development
U.S. Department of Labor
U.S. Department of Energy
Housing Authority of Bowling Green
Kentucky Housing Corporation
Community Action of Southern Kentucky, Inc.
Barren River Area Development District
Workforce Investment Board

Project Description:

The Housing Authority of Bowling Green (population 56,000) used funding from the American Recovery and Reinvestment Act to replace more than 2,000 old, inefficient windows in its public housing units with new energy-efficient windows. The Housing Authority, HUD, and the Kentucky Housing Corporation partnered with Green the Block, the local Workforce Investment Board, and the local Community Action Agency to connect low-income families with green jobs and environmental education opportunities. Green the Block, a partnership between Green for All and the Hip Hop Caucus, aims to ensure that low-income communities, particularly communities of color, participate in and have a voice in the clean energy economy.

Greening the Block in Bowling Green

Livability Principles Addressed:

Promote equitable, affordable housing: Energy-efficient retrofits of public housing units help keep energy costs low, making these homes more affordable for low-income residents of Bowling Green.

Enhance economic competitiveness: Thirty percent more Bowling Green residents have enrolled in colleges, technical trade schools, and other post-high school educational programs since 2008. Training for green jobs improves the local workforce's competitiveness.

Support existing communities: The retrofit of public housing and other homes helps local families stay in their communities.

Coordinate and leverage federal policies and investment: HUD funding to retrofit public housing units not only preserved affordable housing, but also created a larger local market for weatherization services, taking better advantage of Department of Labor programs that train residents for these jobs. These programs also work in tandem with the Department of Energy's weatherization assistance to low-income families, providing even more jobs for these trained workers.

Photos courtesy of HUD and the Housing Authority of Bowling Green

For more information about this project, contact:

Krista Mills
HUD Louisville
502-618-8140
Krista.Mills@hud.gov

Federal Support for Sustainable Rural Communities
Case Studies

U.S. Department
of Transportation
**Federal Transit
Administration**

Downeast Transportation and Island Explorer

Photos, left to right: Downeast Transportation, Volpe Center, Downeast Transportation

Location: Hancock County, Maine

Focus: Rural transit

Funding:
FTA rural and job access funds
National Park Service
L.L.Bean and other local businesses
Municipalities

Partners:
Federal Transit Administration
National Park Service
Maine Department of Transportation
Friends of Acadia
L.L.Bean
Jackson Laboratories
Communities in Hancock County

Project Description:

Downeast Transportation, a private non-profit agency, partners with public and private entities to provide seasonal and year-round transportation services. The 12 Downeast routes connect the towns of Bangor, Bar Harbor, Blue Hill, Ellsworth, and Southwest Harbor. Residents and visitors rely on the service to access jobs, shopping, ferry terminals, trails, and recreation. All transit vehicles carry bicycles, further expanding the range of destinations reachable by transit.

Downeast Transportation runs two primary services—commuter access to major employers such as Jackson Laboratories in Bar Harbor and the Island Explorer shuttle system on Mount Desert Island. Downeast routes bring employees from as far as 60 miles away to Jackson Laboratories and nearby businesses and support multiple shifts. Downeast helps employers create transit-friendly shifts so employees can ride transit to work. Downeast leverages FTA jobs access funds to help provide other transit service in the county at off-peak times. To support riders with special needs, Downeast serves passengers at destinations up to three-quarters of a mile from the fixed-route service at no additional fee.

Through a partnership with the National Park Service, L.L.Bean, and local businesses, Downeast Transportation also operates the seasonal Island Explorer shuttle on Mount Desert Island. The eight routes serving Acadia National Park and the town of Bar Harbor carried more than 400,000 people in 2010. The service provides access to a variety of destinations, reducing pollution and traffic on congested roads. The system enhances the visitor experience by using intelligent transportation systems that provide real-time service information. Transit vehicles include bicycle racks and trailers to support longer trips.

Downeast Transportation and Island Explorer

Community Outreach:

Downeast Transportation began as support for the "Meals for Me" program, bringing seniors to meals, community centers, shopping, and medical appointments. It has evolved to provide access to a range of services to meet community needs.

Downeast is planning a new Welcome Center in Trenton to include offices, vehicle storage and maintenance, an intermodal transit facility, and an Acadia Gateway Visitor Center. This will provide better access to the region for inter-city and day visitors.

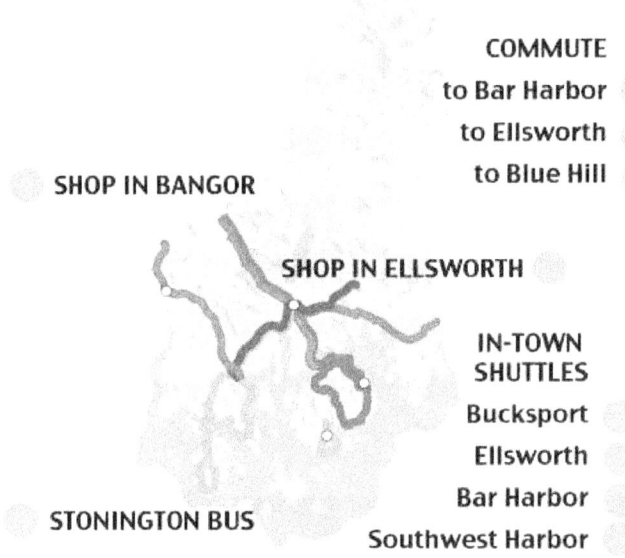

Image: Downeast Transportation

Livability Principles Addressed:

Provide more transportation choices: Downeast's services provide visitors, residents, and commuters with additional travel options. Bicycle racks and trailers on vehicles further expand the range of destinations accessible by transit.

Enhance economic competitiveness: Commuter services accommodate multiple shifts, providing flexibility to support both employers and employees. Access to businesses and tourist destinations supports the local economy.

Support existing communities: Hancock County residents rely on Downeast Transportation services to commute to work, access nearby shopping, and travel to neighboring towns. The seasonal Island Explorer helps to maintain community character by reducing traffic congestion and pollution and supporting the local community.

Coordinate and leverage federal policies and investment: Downeast has partnered with public and private entities, securing long-term funding support from the NPS and L.L. Bean for the Island Explorer service. The agency also creatively combines federal funding sources to better support year-round access to jobs and services throughout Hancock County.

For more information about this project, contact:

Paul Murphy
Downeast Transportation,
General Manager
Paul@ExploreAcadia.com

Peter Butler
Director, Planning and Program Dev.
Federal Transit Administration
Peter.Butler@dot.gov

Prepared for FTA by the U.S. DOT Volpe National Transportation Systems Center This case study, and others related to Livable and Sustainable Communities, is available at: http://fta.dot.gov/publications/publications_10991.html

Federal Support for Sustainable Rural Communities
Case Studies

Opportunity Link: Making Connections with Transit

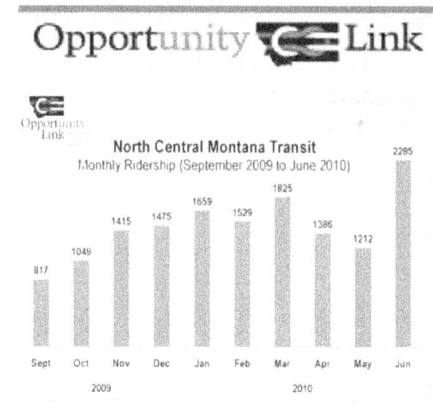

Photos courtesy of Opportunity Link

Location: North central Montana

Focus: Rural and tribal transportation

Partners:
U.S. Department of Transportation

Funding:
Government, business, social service organizations, and educational institutions

Project Description:

With the nearest metropolitan area over 100 miles away, residents of north central Montana lacked access to jobs, educational opportunities, medical case, shopping, and other needed destinations and services. In response to this challenge, Opportunity Link, a non-profit organization that aims to reduce poverty, engaged a broad range of community stakeholders in a regional planning process in 2007. This planning effort convened an unprecedented partnership of government, businesses, and educational institutions from remote tribal and rural communities to explore public transit options. The outcome was the creation of four new rural transit systems: North Central Montana Transit in Hill and Blaine Counties, Fort Belknap Transit Service at Fort Belknap Indian Community, Rocky Boy Transit at the Chippewa Cree Tribe's Rocky Boy Reservation, and Northern Transit Interlocal serving Toole, Pondera, and Teton Counties. Each is designed to respond to the most pressing transportation needs of low-income residents as identified through a needs assessment.

Opportunity Link: Making Connections with Transit

Livability Principles Addressed:

Provide more transportation choices: Opportunity Link enhances transportation options for rural residents, which is particularly important for those who do not have access to private vehicles.

Enhance economic competitiveness: By linking residents of formerly isolated rural towns and tribal reservations, Opportunity Link provides low-income residents with dependable transportation to employment and schooling, enhancing their ability to obtain and keep good jobs and earn a living.

Support existing communities: Opportunity Link serves existing rural towns and neighborhoods, helping to keep them viable places to live.

Coordinate and leverage federal policies and investment: Opportunity Link coordinated funding contributions from government, businesses, social service organizations, and educational institutions, demonstrating an ability to leverage private, local, and federal funds to develop and operate public transportation services.

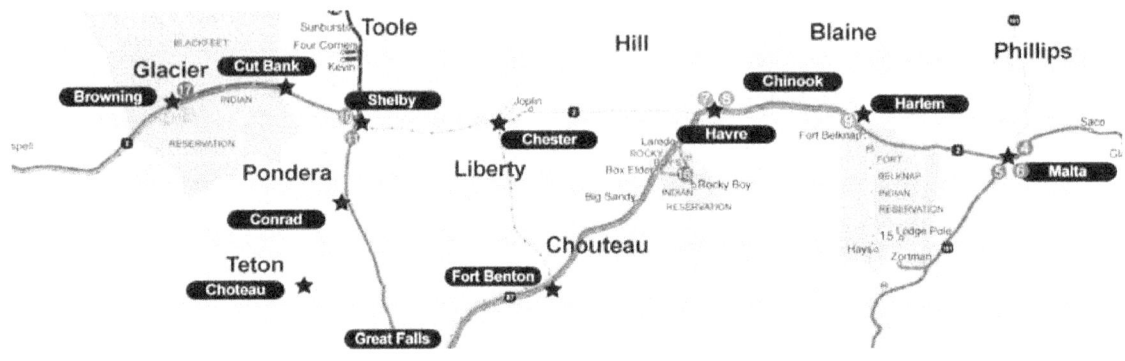

Image: Opportunity Link

For more information about this project, see:

http://www.opportunitylinkmt.org/downloads.php

Federal Support for Sustainable Rural Communities
Case Studies

Maupin Market: Modernizing a Small Town Grocery Store

Photos courtesy of Dennis Ross, Mayor of Maupin

Location: Maupin, Oregon

Focus: Modernizing a rural downtown grocery store

Partners:
U.S. Department of Agriculture – Rural Development
City of Maupin

Funding:
Local bank: $400,000
Small Business Administration: $279,000
Business owner equity: $100,000
Mid-Columbia Economic Development District loan (via USDA-RD Intermediary Relending Program): $100,000

Project Description:

The town of Maupin, Oregon, population 411, is located along the Deschutes River, one of the nation's prime fishing and whitewater rafting streams, and is the most popular destination point along the river. Just 90 minutes from the Portland metropolitan area, Maupin is both a recreation hub and a business service center for the surrounding rural area. Once a typical Oregon mill town, the economy began the transition away from timber in the 1990s and is now dominated by the outdoor recreation market.

When the town's only grocery store was considering closing after 90 years, which would have meant residents would have to travel 40 miles to the next closest grocery store, a vacation homeowner decided to step up and buy the business. After initially considering constructing a new building on the edge of town, the business owner, with encouragement from city leaders, instead opted to purchase and renovate the existing store, which is in the midst of the town's small business district.

A combination of private-sector and federal agency loans helped bring this business proposal to fruition. The result is a completely remodeled building on Maupin's main street, bringing new vitality to its downtown while maintaining a critical community service.

Maupin Market: Modernizing a Small Town Grocery Store

Community Outreach:

In 2005, Maupin's leaders held community meetings to develop a strategic plan. The plan called for reinvestment in the town's historic center where infrastructure already existed and for better walking conditions for the town's many senior residents.

Livability Principles Addressed:

Enhance economic competitiveness: Maintaining a full-service market on the main street keeps residents and visitors coming into downtown Maupin, enhancing the viability of other local businesses and the attractiveness of the town as a place to live and vacation.

Support existing communities: Choosing a centrally located site for the reopened grocery store over a more remote site takes advantage of existing infrastructure and saves public resources that might be required to develop a new site.

Coordinate and leverage federal policies and investment: After typical private loans for business and construction were secured, an additional USDA-RD loan was essential for covering up-front inventory costs.

Value communities and neighborhoods: Keeping the newly opened Maupin Market on Deschutes Avenue, the town's main street, provides an easily accessible service that strengthens the historic town center.

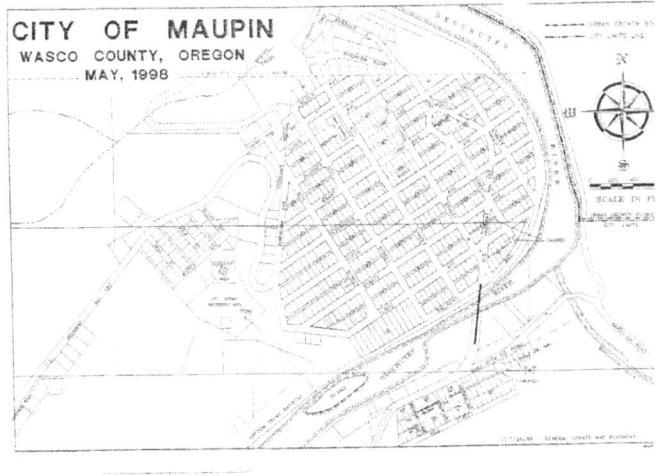

Images, top to bottom: USDA, City of Maupin

<table>
<tr><td>

For more information about this project, contact:
Chris Beck, USDA,
Chris.Beck@osec.usda.gov

</td></tr>
</table>

Federal Support for Sustainable Rural Communities
Case Studies

Howard: Bringing Green Opportunities and Strengthening Downtown

Images courtesy of Rural Learning Center

Location: Howard, South Dakota

Focus: Economic development, main street revitalization, green energy

Funding:
USDA American Recovery and Reinvestment Act: $3,200,000 guaranteed loan
USDA Rural Economic Development Loan Program: $740,000 loan and $300,000 grant

Northeast South Dakota Economic Corporation: $150,000
Grow South Dakota revolving loan fund: $100,000

Partners:
U.S. Department of Agriculture – Rural Development
City of Howard, Miner County
Heartland Consumer Power District
Howard Industries
Citi Foundation

Project Description:

In an effort to spur economic development, Miner County leaders developed a plan for a rural learning center that, in addition to being a community gathering place, will train rural residents on new economic opportunities in rural South Dakota.

In the past 70 years, Miner County has seen a 40 percent decline in farms and businesses and has recently begun diversifying its economy with non-agricultural industries. As a strategy to reduce energy costs and create more business opportunities, the community is developing renewable energy industries. Wind turbines are dotting the landscape, and other green projects are underway.

In 2001, Miner County's efforts were given a boost when the Northwest Area Foundation committed to invest $5.8 million in the Miner County Community Revitalization program over a ten-year period. This support, and the increased revitalization activity it generated, eventually led to a proposal for a Rural Learning Center. Although 40 acres on the outskirts of town were offered free of cost for the center, the project's leaders believed that a more central location on Main Street, while more expensive, would bring more benefits to the community and the local economy. With its location in the heart of Howard, a town of 1,071 people, and its goal of LEED platinum building certification, the Maroney Rural Learning Center demonstrates the community's commitment to downtown revitalization and the new green economy.

Community Outreach:

In the 1990s, public concern over the future of Miner County eventually led to the establishment of Miner County Community Revitalization, a predecessor organization to the Maroney Rural Learning Center. Rural Learning Center leaders engaged the community in a public process that ultimately resulted in the adoption and implementation of a strategic vision and plan for Miner County.

Livability Principles Addressed:

Enhance economic competitiveness: Establishment of the Maroney Rural Learning Center and related convention facilities will enhance business opportunities in Howard. In addition to new employees, an influx of visitors will create new markets for Main Street businesses. Increased sales tax revenues will support future infrastructure investments in town.

Support existing communities: Using a centrally located site takes advantage of existing community infrastructure and saves public resources that might be required for developing a new site outside of town.

Coordinate and leverage federal policies and investment: The Maroney Rural Learning Center combines funding from various federal, state, and private sources to support the community's goals of promoting economic development, revitalizing its Main Street, and saving money on energy costs.

Value communities and neighborhoods: The Maroney Rural Learning Center's location in the heart of Howard strengthens and builds off of Howard's distinctive business district and adjacent neighborhoods.

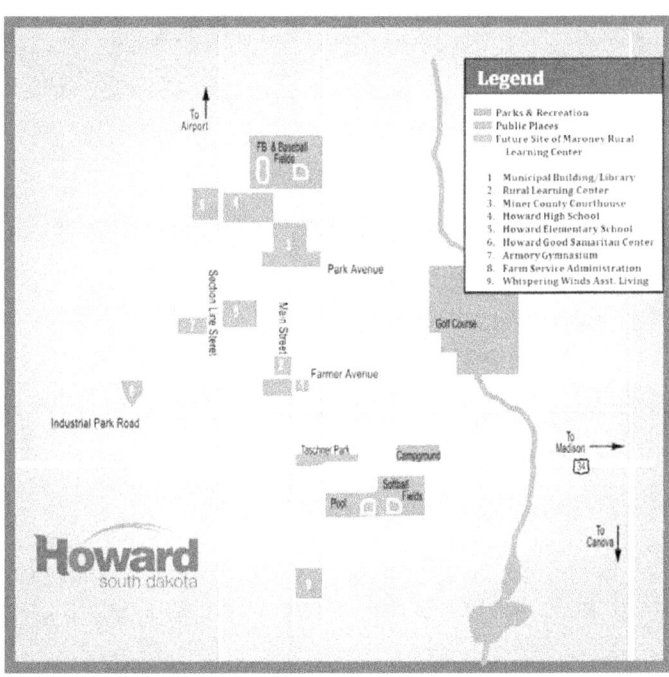

Images, top to bottom: USDA, City of Howard

For more information about this project, contact:
Joe Bartmann, Rural Learning Center, 605-772-5153, joe@rurallearningcenter.org

Federal Support for Sustainable Rural Communities
Case Studies

Renewing the Community in Thunder Valley

Photos courtesy of Oyate Omniciyé

Location: Pine Ridge Reservation (Oglala Lakota tribe), South Dakota

Focus: Community building and preservation, economic development

Funding:
HUD: $996,100

Partners:
U.S. Department of Housing and Urban Development
Thunder Valley Community Development Corporation
BNIM, Inc.
inNative
Tribal President's Office

Project Description:

With support from HUD's Sustainable Communities Regional Planning Grant program, the Thunder Valley Community Development Corportation and the Oglala Lakota Tribe are leading an effort to develop a Regional Plan for Sustainable Development in their remote area of southwestern South Dakota covering the Pine Ridge Indian Reservation. The reservation has no active planning department, and this will be its first comprehensive and integrated plan.

Oglala Lakota community members hope the plan will help address the long standing challenges they have faced. Economic opportunity is scarce on the reservation, and the unemployment rate is near 47 percent. Additionally, many low-income families cannot find affordable housing and end up leaving the reservation, moving into crowded households with family members, or joining the reservation's homeless population. The Tribal Council estimates that at least 4,000 new homes are needed. Residents also have limited access to services. There is one medium-sized grocery store on the reservation and no bank. 80 percent of the money spent on retail is spent outside the reservation.

HUD funding is enabling the reservation's Housing Authority, Environmental Protection Program, Chamber of Commerce, Health Administration, and other agencies to collaborate on the development of the regional plan. These local partners are supported by BNIM, Inc., a private firm with experience working with rural and tribal communities. BNIM has placed a full-time staff member on the reservation in order to better understand the priorities and visions of the community.

Renewing the Community in Thunder Valley

Livability Principles Addressed:

Promote equitable, affordable housing: To address the reservation's housing shortage, the consortium will conduct a housing and market analysis that takes into account employment patterns, regional industries, tourism trends, and local skills.

Enhance economic competitiveness: The consortium will identify sectors that could be competitive in the regional economy, clarify business regulations to enhance access to capital, and increase worker opportunities through skills training, especially for the youth population.

Support existing communities: Planning that strengthens tribal culture, rebuilds the spiritual fibers of society, and preserves the unique knowledge of the Lakota is at the heart of the consortium's work. The Oglala Lakota language and cultural traditions are the foundations of any planning process, whether it is for economic development, housing, or transportation. Planning will draw on Lakota values and focus on self-sufficiency.

Coordinate and leverage federal policies and investment: The consortium has recognized the importance of leveraging this HUD grant to obtain resources from other agencies. They have already reached out to various agencies' field offices to help identify applicable funding sources for projects on the reservation.

Value communities and neighborhoods: The plan seeks to continue the healing and strengthening of the Oglala Lakota people by bolstering identity and opportunity through the unique perspective of Lakota knowledge, culture, and language.

Photos courtesy of Oyate Omniciyé

For more information about this project, contact:
Dwayne Marsh, HUD Office of Sustainable Housing and Communities
Dwayne.S.Marsh@hud.gov

Community Involvement:

Residents will be involved in all stages of the planning process. Members of the Thunder Valley Community Development Corporation attend community meetings, and a steering committee will be appointed to make decisions about the plan. The consortium has changed its name to Oyate Omniciyé, a Lakota phrase meaning "Circle Meetings of the People," to better reflect the cultural community-based process they will use in developing the plan.

Youth involvement is particularly important to ensure broad community support for the plan. Project staff are using Facebook, where the project has over 300 fans, and also appearing on radio shows to discuss what sustainable community planning means for the Pine Ridge Reservation.

The planning process being led by Oyate Omniciyé is a model for tribes around the country. The work being done by the Oglala Lakota is now nationally known in tribal circles, and they are sharing their experience and encouraging other tribes to develop collaborative planning efforts and applications for the Sustainable Communities Regional Planning Grant program.

Federal Support for Sustainable Rural Communities
Case Studies

Tennessee Intercity Bus Program: Providing Rural Opportunity

Photo courtesy of the Tennessee Department of Transportation

Location: Rural Tennessee

Focus: Rural transit, access to opportunity

Partners:
U.S. Department of Transportation

Funding:
DOT: $3.1 million

Project Description:

In response to growing public demand, the State of Tennessee implemented the Tennessee Intercity Bus Demonstration Program in 2008. The intercity bus program provides Tennesseans in rural areas with reliable daily access to health care, jobs, schools, and other destinations in the state's metropolitan areas. It offers the Amish community and other rural residents more transportation choice and greater access to opportunities and services.

Tennessee Intercity Bus Program: Providing Rural Opportunity

Livability Principles Addressed:

Provide more transportation choices: This program increases transportation choices, particularly for citizens in rural areas who do not drive.

Enhance economic competitiveness: The intercity bus program boosts the economic competitiveness of rural residents by connecting them to employment and education in metropolitan areas, allowing them to make a living and build their skills.

Support existing communities: The bus routes serve many rural towns and village centers, making them more attractive to new and existing residents.

Coordinate and leverage federal policies and investment: The project was implemented through coordination among many stakeholders, including the Nashville metropolitan planning organization and local city and county officials from 15 counties.

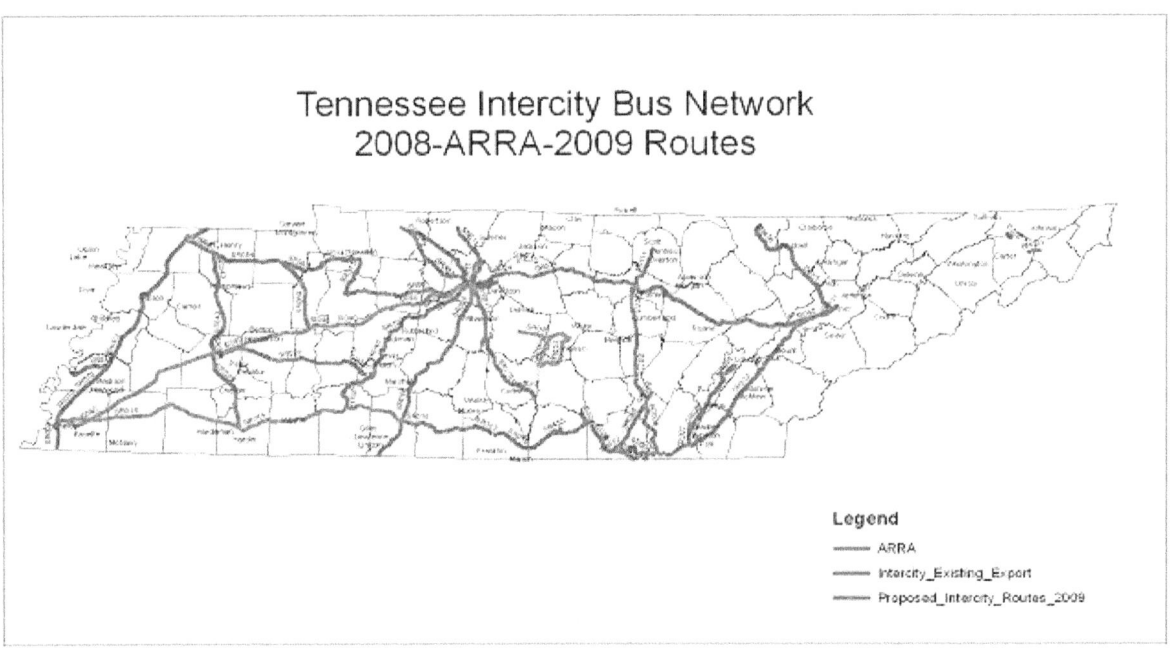

Image courtesy of the Tennessee Department of Transportation

For more information about this project, see:
http://www.fta.dot.gov/publications/about_FTA_10999.html

Federal Support for Sustainable Rural Communities
Case Studies

Ranson-Charles Town Corridor Revitalization

Photos courtesy of City of Ranson

Location: Ranson and Charles Town, West Virginia

Focus: Corridor revitalization, complete streets, green infrastructure

Funding:
HUD: $271,500
DOT: $708,500
EPA: $620,000
West Virginia Department of Transportation: $174,000

Partners:
U.S. Department of Housing and Urban Development
U.S. Department of Transportation
U.S. Environmental Protection Agency
West Virginia Department of Transportation

Project Description:

Ranson and its partner community, Charles Town, launched the Commerce Corridor Initiative in 1999 to revitalize a corridor that runs through their nearby downtowns. More recently, the two cities launched a complete street/green corridor revitalization project for a boulevard that intersects the commerce corridor. The two cities have received almost $2 million in federal and state grants to help support the two projects.

The cities of Ranson and Charles Town, with populations of 4,000 and 4,274 respectively, are combating the effects of recent manufacturing and other facility closures, as well as increasing growth pressures from the Baltimore-Washington metropolitan area. In the past several years, Ranson alone has lost more than 1,500 jobs, leaving the community with contaminated, idled, and vacant sites and a downtown area that was increasingly falling into disrepair. At the same time, the Jefferson County population has steadily grown, transforming the rural area into an exurb, with few strategies to guide the growth. More than 25 percent of nearby residents live below the poverty line. Furthermore, over the years, the main corridor between the two towns has turned into an auto-dominated roadway that is unsafe for pedestrians and bicyclists due to the lack of sidewalks and safe cross-walks. To combat the disrepair, preserve the character and history of the towns, and promote economic development, the cities of Ranson and Charles Town have developed two corridor revitalization initiatives.

Ranson-Charles Town Corridor Revitalization

In 1999, Ranson and Charles Town launched the Commerce Corridor Initiative with the goal of creating a high-tech commerce corridor, complete with high-tech and commercial offices, retail, entertainment amenities, infill housing, parks and recreational areas, and government facilities. It will be located on a corridor of formerly vacant properties, many of which are brownfield sites, that runs through the cities' downtowns. The cities put together a Commerce Corridor Council to advise the effort, and several city resolutions and agreements advanced the initiative.

Ranson and Charles Town received federal financial support from EPA to help fund brownfields assessment, the creation of clean up and reuse plans for certain sites, and community outreach efforts. EPA awarded Ranson and Charles Town Brownfields Assessment Grants in 2001, 2004, and 2006. More recently, they received an EPA Brownfields Area-Wide Planning Pilot Program grant in 2010 to help them develop an area-wide plan. The area-wide plan will help Ranson and Charles Town prioritize brownfields site assessment and clean up and develop site-specific reuse plans based on community input.

Ranson also received EPA Sustainable Communities Building Blocks assistance to help the city identify and address common barriers to smart growth implementation and help them develop in a way that is environmentally and economically sustainable.

Ranson and Charles Town have already made some progress in their goal of economic revitalization in the commerce corridor. For example, Powhatan Place is a new, mixed-use, infill development that is designed to meet LEED for Neighborhood Development standards and is located on the site of a former foundry. The development includes a mix of housing types, stores, public spaces, recreation areas, trails, and green infrastructure elements to manage stormwater runoff.

In a related effort, Ranson and Charles Town recently received DOT TIGER II and state funding to redesign the Fairfax Boulevard-George Street corridor, which runs through both downtowns and intersects the commerce corridor. This boulevard will become a "complete street," a street designed and operated to enable safe access for all users, including pedestrians, bicyclists, motorists, and transit riders. It will also incorporate green infrastructure elements. Additionally, these funds will support the redesign of an adjacent historic building as a new regional commuter center in downtown Charles Town, which will provide residents and workers with access to regional rail and bus transit.

A HUD Community Challenge Planning Grant will help Ranson create a new zoning code to foster environmentally and economically sustainable community development. The code will link a green downtown overlay district with a new zoning approach for the city's undeveloped, outlying areas. This effort will help Ranson and Charles Town create a vibrant, mixed-use corridor that

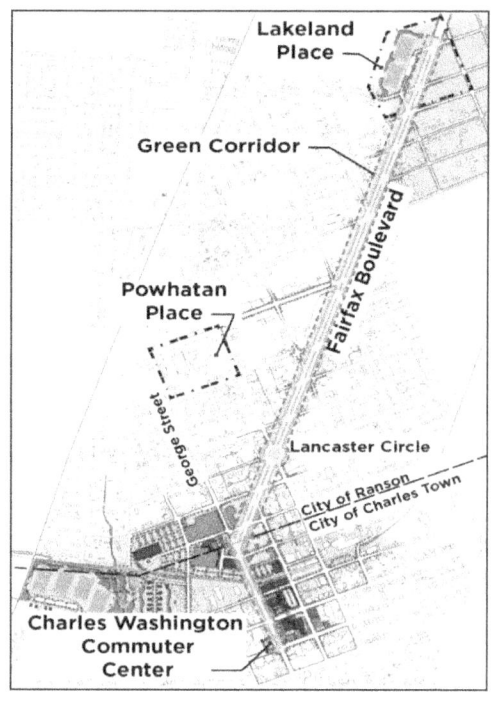

Image courtesy of City of Ranson

is transit-oriented, walkable, and bikeable and that provides access to regional job centers and community facilities.

Stakeholder involvement has been important in many of the assessment, planning, and revitalization efforts and will continue to be vital. For example, city of Ranson staff and consultants will work with community members to determine the vision for the complete street design and the preliminary sketches for the commuter center. In addition, they will host a two-day public workshop on zoning codes and a seven-day workshop to produce illustrative plans for mixed-use corridor redevelopment and the downtown green overlay district.

For more information about this project, contact:
Sunaree Marshall, HUD, 202-402-6011, sunaree.k.marshall@hud.gov

Appendix B: Partnership for Sustainable Communities Charge to the Rural Work Group
August 2010

Rural America makes up about 16 percent of the country's population and covers 75 percent of the land area. Rural America includes towns and small cities, as well as working lands, farms, prairies, forests, and rangelands. Challenges in rural communities include a declining agriculture, natural resources economy, and the smaller scale manufacturing economy (especially in rural areas removed from metro-areas and natural amenities), farmland and natural-area land conversion in high amenity or areas adjacent to metro areas, and the lack of capacity and technical expertise to address these challenges. The Partnership's Livability Principles apply to rural, urban, and suburban communities alike. To rural communities and supporters of the interests of rural communities, the work of the Partnership and the benefits to rural America need to be made clear.

Work Group Charge:

1. Create a concise framework describing how the Partnership's Livability Principles support rural communities across the country. The framework may address policy efforts of the Partnership and the member agencies as well as messaging and framing the work. One question that likely needs to be answered is "How do the Livability Principles increase economic opportunity and increase quality of life for all rural Americans."

2. As a first step towards identifying important policy leverage points, describe current efforts in Partnership agencies (across the Administration?) that are focused on the built environment in rural communities. These could include programs, policies, rules, and statutes within in agencies and departments. Particular emphasis could be placed on economic development opportunities and existing policies, etc, that leverage other investments.

3. Identify a set of case studies showing rural communities successfully using smart growth/sustainable community approaches to achieve better economic, environment, community, and public health outcomes.

4. Identify stakeholder groups – locally in communities and nationally – NGOs, trade associations, and other groups that do work around the built environment in rural communities. Put forward a method for ensuring the variety of rural voices are heard in the context of the work of the Partnership.

5. Propose an approach to outcome-based livability performance measures for rural communities / regions.

Appendix C: Rural Work Group Members

Partnership for Sustainable Communities Rural Work Group Contact List			
Agency	First Name	Last Name	Email Address
USDA	Chris	Beck	Chris.Beck@osec.usda.gov
USDA	Megan	Bolin	Margaret.Bolin@osec.usda.gov
USDA	Annie	Goode	Annie.Goode@wdc.usda.gov
USDA	Doug	O'Brien	Doug.O'Brien@osec.usda.gov
USDA	David	Sears	David.Sears@wdc.usda.gov
HUD	Daniel	Lurie	Daniel.B.Lurie@hud.gov
HUD	Stewart	Sarkozy-Banoczy	Stewart.G.Sarkozy-Banoczy@hud.gov
HUD	Carrie	Schuettpelz	Carrie.A.Schuettpelz@hud.gov
HUD	Rachel	Thornton	Rachel.J.Thornton@hud.gov
HUD	Mariia	Zimmerman	Mariia.Zimmerman@hud.gov
DOT	Eric	Beightel	Eric.Beightel@dot.gov
DOT	Fred	Bowers	Frederick.Bowers@dot.gov
DOT	James	Cheatham	james.cheatham@dot.gov
DOT	Mary Martha	Churchman	MaryMartha.Churchman@dot.gov
DOT	Charlie	Goodman	Charles.Goodman@dot.gov
DOT	Aung	Gye	Aung.Gye@dot.gov
DOT	Bryna	Helfer	bryna.helfer@dot.gov
DOT	Linda	Lawson	Linda.Lawson@dot.gov
DOT	Yuh Wen	Ling	Yuhwen.ling@dot.gov
DOT	Kate	Mattice	Katherine.Mattice@dot.gov
DOT	Kenneth	Petty	Kenneth.Petty@dot.gov
DOT	Jeffrey	Reczek	jeffrey.reczek@dot.gov
DOT	Gabe	Rousseau	Gabe.Rousseau@dot.gov
DOT	Rebecca	Searl	Rebecca.searl@dot.gov
DOT	John	Sprowls	John.Sprowls@dot.gov
DOT	Lorna	Wilson	lorna.wilson@dot.gov
EPA	Stephanie	Bertaina	Bertaina.Stephanie@epa.gov
EPA	Ann	Carroll	Carroll.Ann@epa.gov
EPA	Matthew	Dalbey	Dalbey.Matthew@epa.gov
EPA	Rabi	Kieber	Kieber.Rabi@epa.gov
EPA	Megan	McConville	Mcconville.Megan@epa.gov
EPA	Carolyn	Mulvihill	Mulvihill.Carolyn@epa.gov
EPA	Kevin	Ramsey	Ramsey.Kevin@epa.gov
EPA	Tim	Torma	Torma.Tim@epa.gov
White House	David	Lazarus	David_J._Lazarus@who.eop.gov

.

www.ingramcontent.com/pod-product-compliance
Lightning Source LLC
Chambersburg PA
CBHW080648180526
45168CB00008B/3339